CONSTELLATIONS FOR KIDS

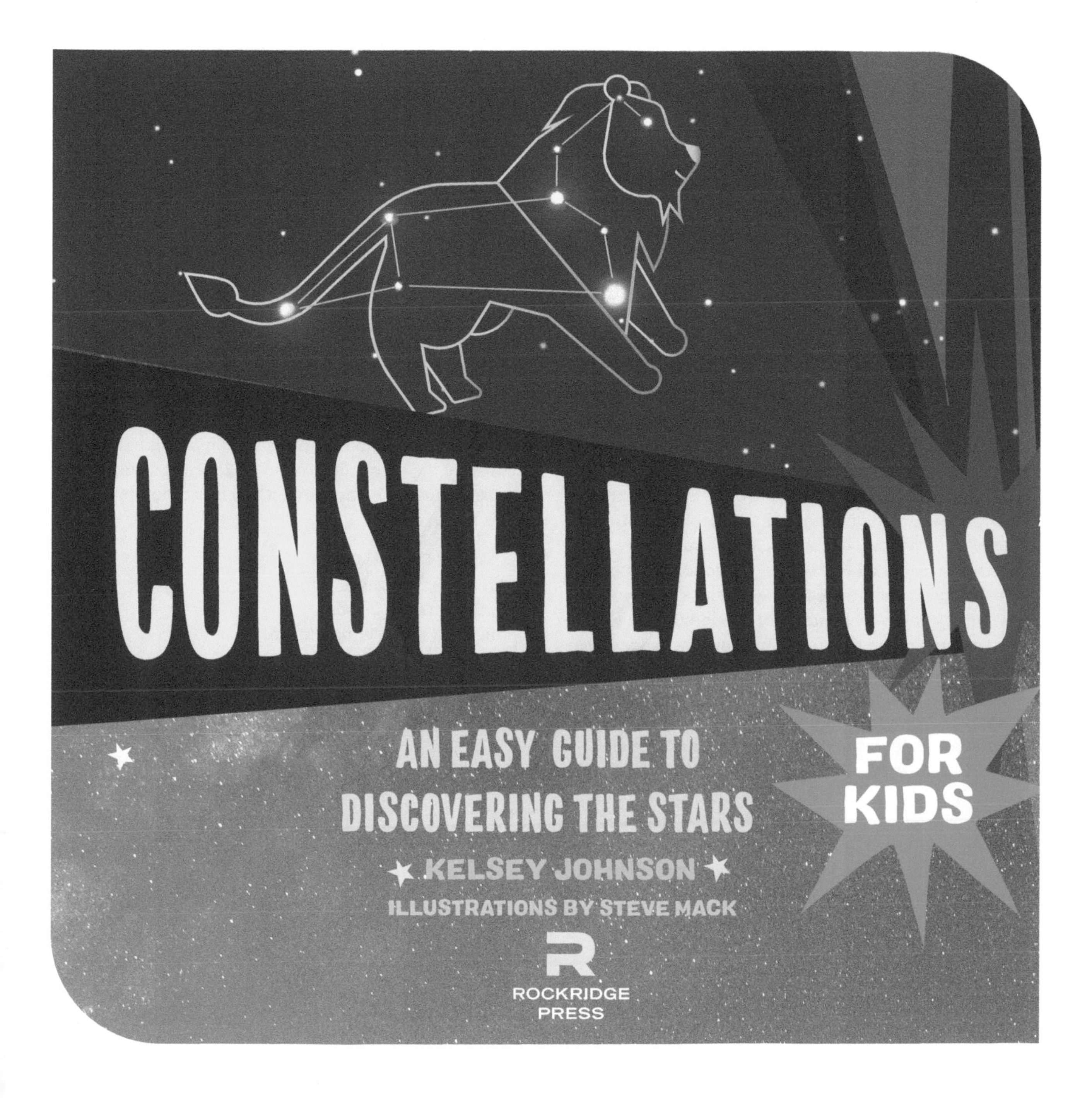
CONSTELLATIONS
AN EASY GUIDE TO
DISCOVERING THE STARS
FOR KIDS
KELSEY JOHNSON
ILLUSTRATIONS BY STEVE MACK
R
ROCKRIDGE
PRESS

Interior and Cover Designer: Jane Archer
Art Producer: Samantha Ulban
Editor: Laura Bryn Sisson
Production Editor: Jenna Dutton

The author used Stellarium (stellarium.org), a free and open-source stargazing program, to create the star maps shown throughout this book.

ISBN: Print 978-1-64611-968-4
eBook 978-1-64611-969-1

R0

To all of the
stargazers
who ponder the
night sky and wonder
what might be.

CONTENTS

★☆☆ Beginner
★★☆ Intermediate
★★★ Advanced

As you read, you'll see tricky words in **bold**. Find out what these words mean in the glossary on page 73.

THE AMAZING UNIVERSE!

ou are part of an amazing universe! People have gazed at the night sky for thousands of years and wondered what they were seeing. Different cultures have seen different pictures in the stars and imagined magical stories. These pictures made of stars in the sky are called **constellations**. We have even used the stars and constellations to help us. For example, Polaris (also called the North Star) has guided many people on journeys. Today very few people know the constellations. Are you ready to begin learning the ancient skill of how to "read" the night sky? Let's get started!

DID YOU KNOW ?

THE STARS YOU CAN SEE CHANGE DURING THE NIGHT!

The Earth is always turning. As it turns, we can see different parts of the sky.

Try This!

Slowly turn around
and around.
3 . . . 2 . . . 1 . . .
Freeze!
What do you see?
Again, slowly turn around and
around. **3 . . . 2 . . . 1 . . .**
Freeze!
Now what do you see?
Do you see all of the same things as before?

Another word for "**turn**" is "**rotate**."

The same thing happens with Earth. The stars you can see in the sky change as the Earth turns.

DID YOU KNOW ?

THE STARS YOU CAN SEE CHANGE WITH THE SEASONS!

Another way to say "circle around" is "orbit."

The Sun is a star that the Earth circles around. During the day, when the Sun is in the sky, it is so bright that we can't see other stars. We have to wait until after the Sun sets.

As the Earth rotates, we can see different stars in the night sky. But every year we can see the same stars at the same time of year.

About a millon earths can fit inside the sun!

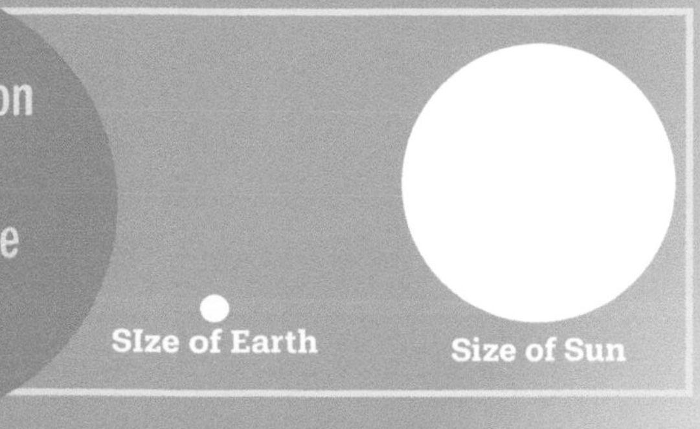

SIze of Earth

Size of Sun

The sun is actually much farther away from the Earth than shown here.

DID YOU KNOW ?

THE STARS YOU CAN SEE DEPEND ON WHERE YOU ARE!

The Earth is shaped like a big ball. There is an imaginary line around the middle of the Earth called the **equator**.

The equator divides the Earth into two halves. Each half is called a **hemisphere**.

You can only see stars that are above where you are on the Earth. If you take a trip to the other hemisphere, you will see different stars!

In this book, we will learn about constellations that can be seen from the Northern Hemisphere.

Another word for "**ball**" is "**sphere**."

"**Hemi**" means "**half**." So "**hemisphere**" means "half of a **sphere**."

WHAT'S A CONSTELLATION?

There are **88** official constellations!

Have you ever done a connect-the-dots puzzle? Constellations are like connect-the-dots puzzles in the sky, where the stars are the dots. People have looked at the stars for thousands of years and imagined connecting them into patterns. These patterns of stars are called constellations.

Many constellations are named after mythical creatures and people.

There are also patterns of stars called **asterisms**. Asterisms are smaller than constellations. Some asterisms, like the Big Dipper, are famous.

Different cultures have seen different patterns in the stars over time. In this book, we will learn about the official constellations used by scientists today.

If you are far away from city lights, you can see about **6,000** stars with your own eyes!

YOU CAN SEE OTHER THINGS IN THE NIGHT SKY, TOO!

When you are looking at the stars, you might see some other objects, too.

The Moon

Sometimes we see the Moon in the night sky. How much of the Moon we see depends on the Sun. Yes, the Sun! Even when it is night on Earth, the Sun is lighting up the Moon. How much of the Moon is lit by the Sun is called the Moon's **phase**. A full moon is the phase when the entire Moon is lit up.

The word "**month**" comes from the word "**moon**," because the Moon takes one month (moon-th) to go around the Earth.

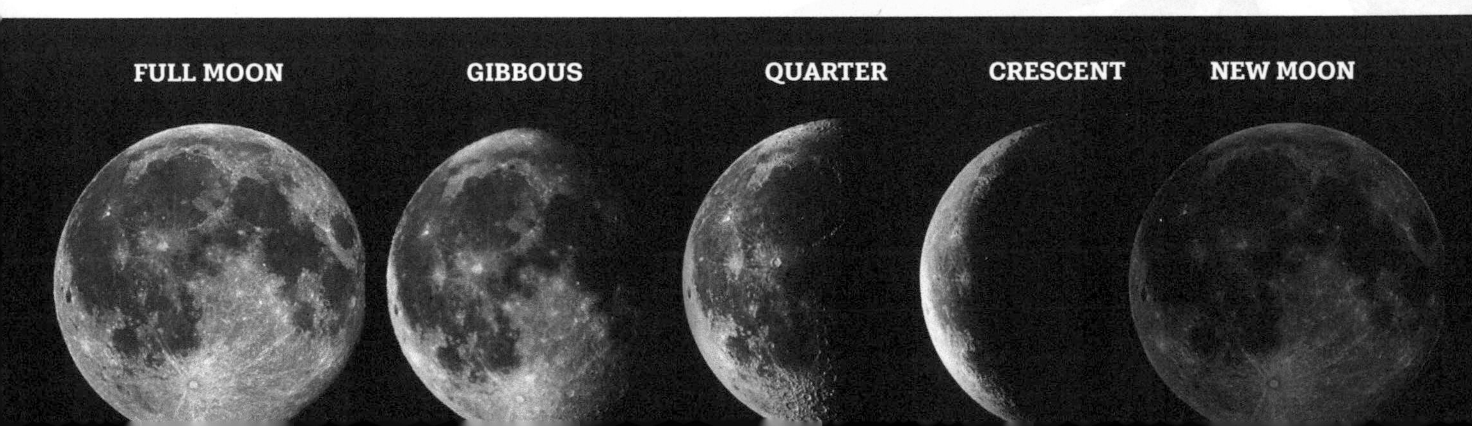

It's harder to see stars during a full moon. To see the faintest stars, it is better to go stargazing during the phases called **new moon** or **crescent moon**.

SATURN

Planets

As the planets **orbit** the Sun, sometimes you can see them in the sky at night. Even though the planets move, we can predict where they will be on any day. See the Resources page in the back of this book to find out how.

The planets Mercury, Venus, Mars, Jupiter, and Saturn are all bright enough to see with your own eyes when they're in the nighttime sky. They look like really bright stars.

Shooting Stars

Shooting stars are not really stars—they are **meteors**. Meteors are pieces of dust or rock from space that are flying through the sky.

Sometimes there are **meteor showers**—lots of meteors! For a calendar of meteor showers, see the Resources page.

Satellites

When you are stargazing, you might see might see something that looks like a faint star moving across the sky. This is a **satellite**. People have sent thousands of satellites into space to orbit the Earth.

Some satellites go all the way around Earth **18** times in one day.

HOW TO USE THIS BOOK

To find constellations, you need to know where to look! As the Earth rotates and orbits the Sun, constellations look like they're at different places in the sky. Constellations in this book are grouped into the best times of year to see them. Each constellation also has a chart that will tell you the direction and how high from the horizon to look in different months at the same time of night.

This book will show you two ways to find constellations:

1. You can use the charts to see where and how high to look in the sky.
2. You can use bright stars as guides to show you where to look.

A Handy Trick

Astronomers tell how far apart things are in the sky using **degrees**. These degrees are not temperatures. Sometimes degrees are shown with the symbol °.

Make a fist and hold it up at arm's length. The width of your fist is about 10 degrees. No matter how old you are, this trick will work. As you grow up your fist will get bigger, but your arm will get longer, too. Now you can tell how far apart things are using your fists!

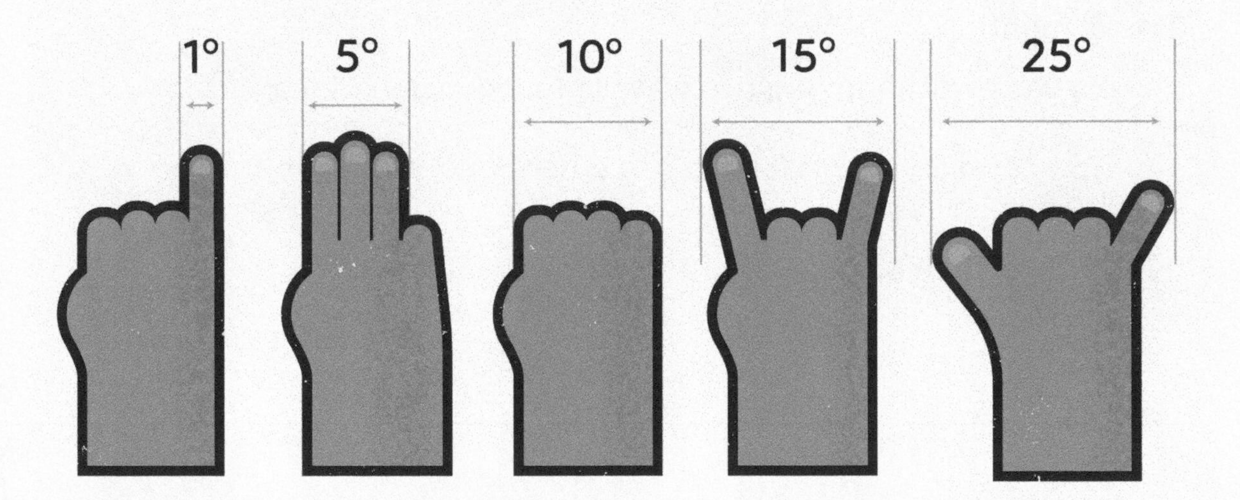

Using Maps of the Sky

A sky map is a picture of the night sky. It helps to use a sky map for the same time and season you are stargazing. Sky maps show you where to look in the sky to find a constellation. You might think that the directions east and west seem backward on a sky map. That is because you need to look at the sky map like you are looking *up* at the night sky.

The constellations in this book are organized into when you can best see them: spring, summer, fall, or winter. At the end of the book, on pages 79–82, you will find a sky map for each season of the year. In the winter, you can see stars in the sky much earlier each night than you can in the summer, because the sun sets earlier. In this book, the winter sky map and constellation charts are set for 8:00 p.m. The spring and fall sky maps and constellation charts are set for 9:00 p.m. The summer sky map and constellation charts are set for 10:00 p.m. Hopefully you can stay up that late!

North, South, East, and West

Another way to find what direction to face is to use a compass. You can use a real compass or a compass app on a cell phone. There are many ways to use a compass, but finding which direction to face is easy. Compasses have a needle that points north. If you turn your body so that you are facing in the direction of the needle, then north is ahead of you, west is on your left, east is on your right, and south is behind you.

Try This!

Use a compass to find out which direction is north. Now, turn to a sky map in this book (pages 79–82), and hold the book above your head. Turn so "north" on the map is really facing north. Now east and west should be in the right directions!

USING BRIGHT STARS AS YOUR GUIDES

Bright stars are like landmarks in the sky. Learning a few bright stars will help you find constellations.

Using star-by-star steps, you can follow a path from a bright star or constellation to another constellation in the sky.

Finding Landmarks in the Sky

Two of the most important landmarks to find in the Northern Hemisphere are the Big Dipper and the North Star. Once you find these, you can use them to find many constellations. Let's use these as an example of how you'll find other constellations in this book.

WHAT DO YOU SEE?

Many different cultures in history have seen a bear shape in the stars of Ursa Major. Other cultures have seen the shape of a wagon, a plow, or even a coffin!

The Big Dipper

The Big Dipper is not actually a constellation: it is an asterism. This pattern of stars is one of the easiest to recognize in the northern sky. The Big Dipper is part of Ursa Major, the big bear.

WHERE TO LOOK FOR THE BIG DIPPER AT 9:00 P.M. IN EACH SEASON				
When to look:	SPRING	SUMMER	FALL	WINTER
Where to look:	NE	NW	N	NW
How high to look:	5 fists (50°)	5 fists (50°)	1 fist (10°)	2 fists (20°)

The Big Dipper is **25°** (2½ fists) from end to end. Use the Big Dipper to test your fist size! See page 10 for more information.

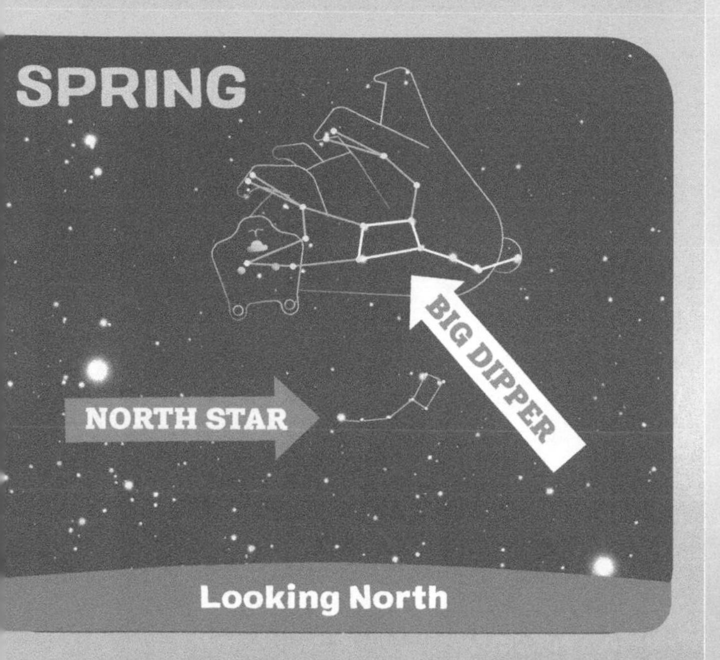

USE THESE PICTURES TO FIND OUT WHERE TO LOOK FOR THE BIG DIPPER IN DIFFERENT SEASONS.

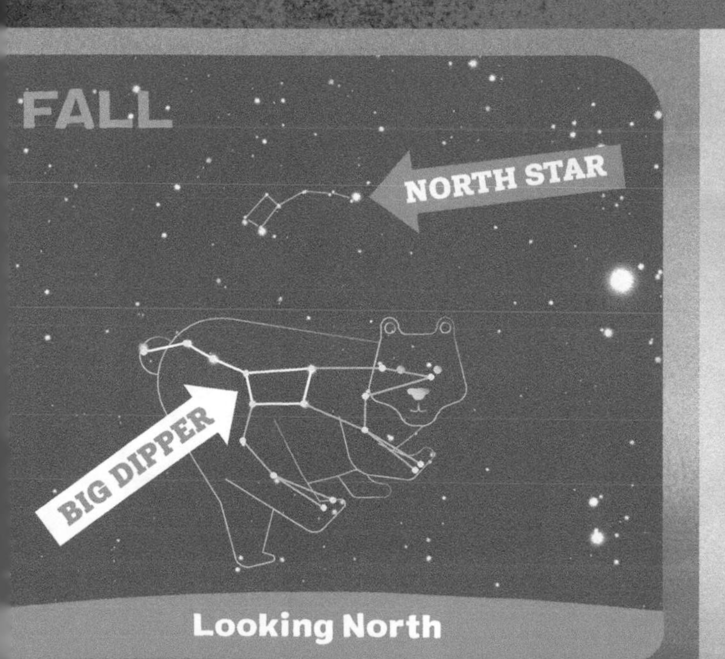

The North Star

The North Star, or *Polaris*, is one of the most famous stars. It is known as the North Star because an imaginary line from the North Pole of the Earth points at it.

As the Earth spins, the other stars in the sky appear to rotate around the **North Star**.

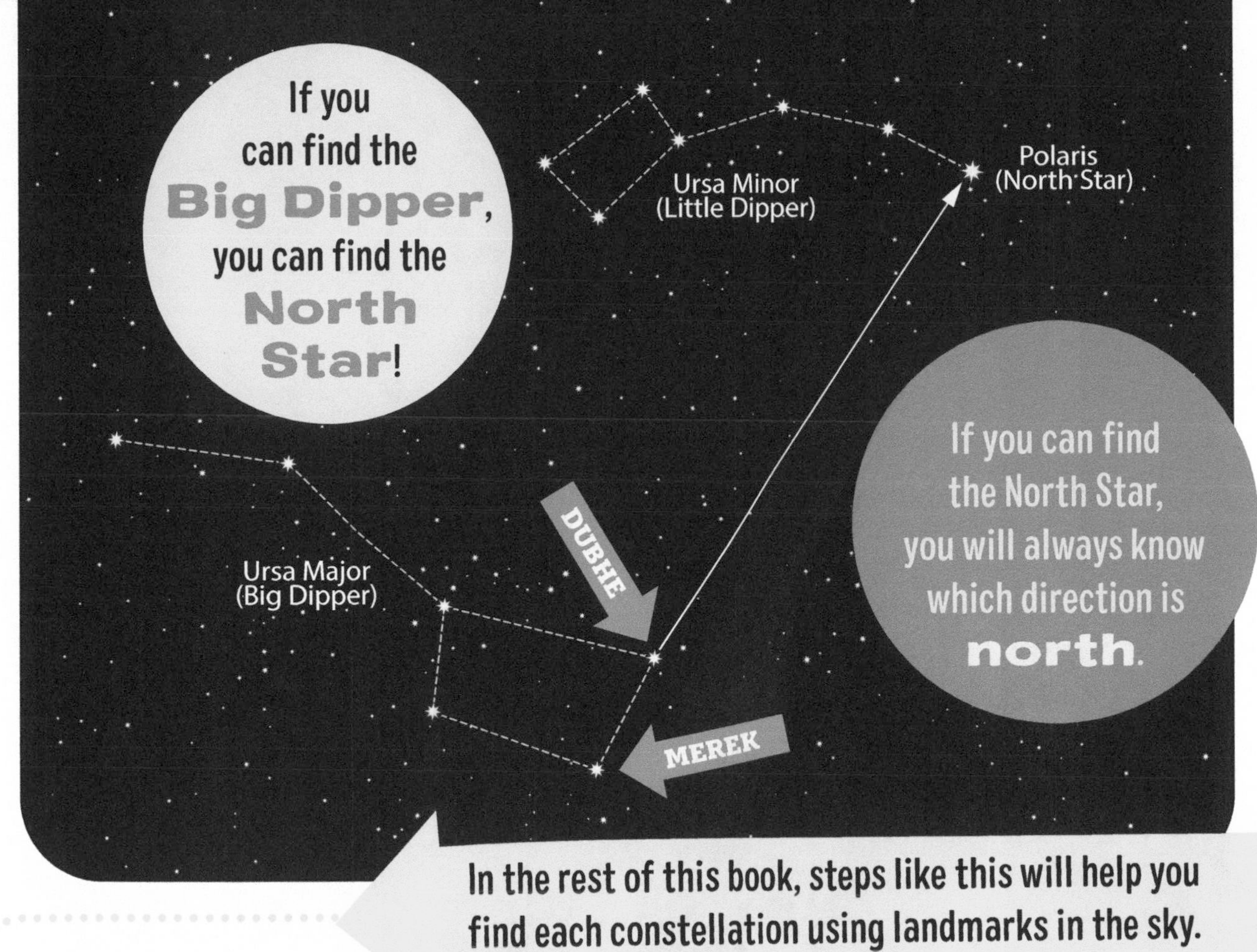

In the rest of this book, steps like this will help you find each constellation using landmarks in the sky.

STAR BY STAR

1. Find the Big Dipper.
2. Locate the two stars on the outside of the "spoon" of the dipper. These two stars, *Merak* and *Dubhe*, are called the **Pointer Stars**.
3. Draw an imaginary line through the Pointer Stars toward the top of the "spoon."
4. Extend that line 30° (three fists) and you will land on the North Star.

CONSTELLATIONS BEST TO SEE IN SPRING

Leo
Boötes
Virgo
Cancer

ZOSMA

ALGIEBA

The constellation Leo has been recognized for **6,000 years**!

Leo

BEGINNER ★☆☆

Leo, the lion. Look for the backward question mark in the sky!

BEST MONTHS TO VIEW AT 9:00 P.M.: MARCH TO MAY			
When to look:	MARCH	APRIL	MAY
Where to look:	E	SE	S
How high to look:	4 fists (40°)	6 fists (60°)	6 fists (60°)

STAR BY STAR

1. Find the Big Dipper (see page 14 to find out how).
2. Identify the Pointer Stars, *Merak* and *Dubhe*, on the side of the Dipper's spoon.
3. Imagine a line connecting the Pointer Stars.
4. Extend that imaginary line out the *bottom* of the spoon.
5. Follow that line about three and a half fists (35°).
6. Your line should land just in front of Leo's tail!

HERCULES AND THE LION

Hercules, son of the Greek god **Zeus**, was given 12 very hard tasks by a powerful king. The first was to kill a dangerous lion that could not be hurt by any weapon. Hercules defeated the lion with his bare hands, and Zeus celebrated the defeat of the lion by placing it in the night sky.

THE INVENTION OF THE WAGON

According to some stories, Boötes invented the wagon and plow. He used them to dig up soil to plant crops. He was placed in the sky to honor his invention.

Boötes

INTERMEDIATE ★★☆

Boötes, the herdsman. To say "Boötes" like an astronomer, you pronounce it *bo-OH-tees*. The main stars in Boötes are shaped like a giant kite. Boötes has one of the brightest stars in the sky, which is named Arcturus.

BEST MONTHS TO VIEW AT 9:00 P.M.: APRIL TO JULY				
When to look:	APRIL	MAY	JUNE	JULY
Where to look:	E	E	SE	SW
How high to look:	2 fists (20°)	5 fists (50°)	7 fists (70°)	6 fists (60°)

STAR BY STAR

1. Find the Big Dipper (see page 14 to find out how).
2. Identify the handle of the Big Dipper.
3. Imagine a line following the curve of the handle.
4. Extend that imaginary line out the end of the handle.
5. Follow that curve for three fists (30°).
6. Your line will land on the brightest star in Boötes, called *Arcturus*.

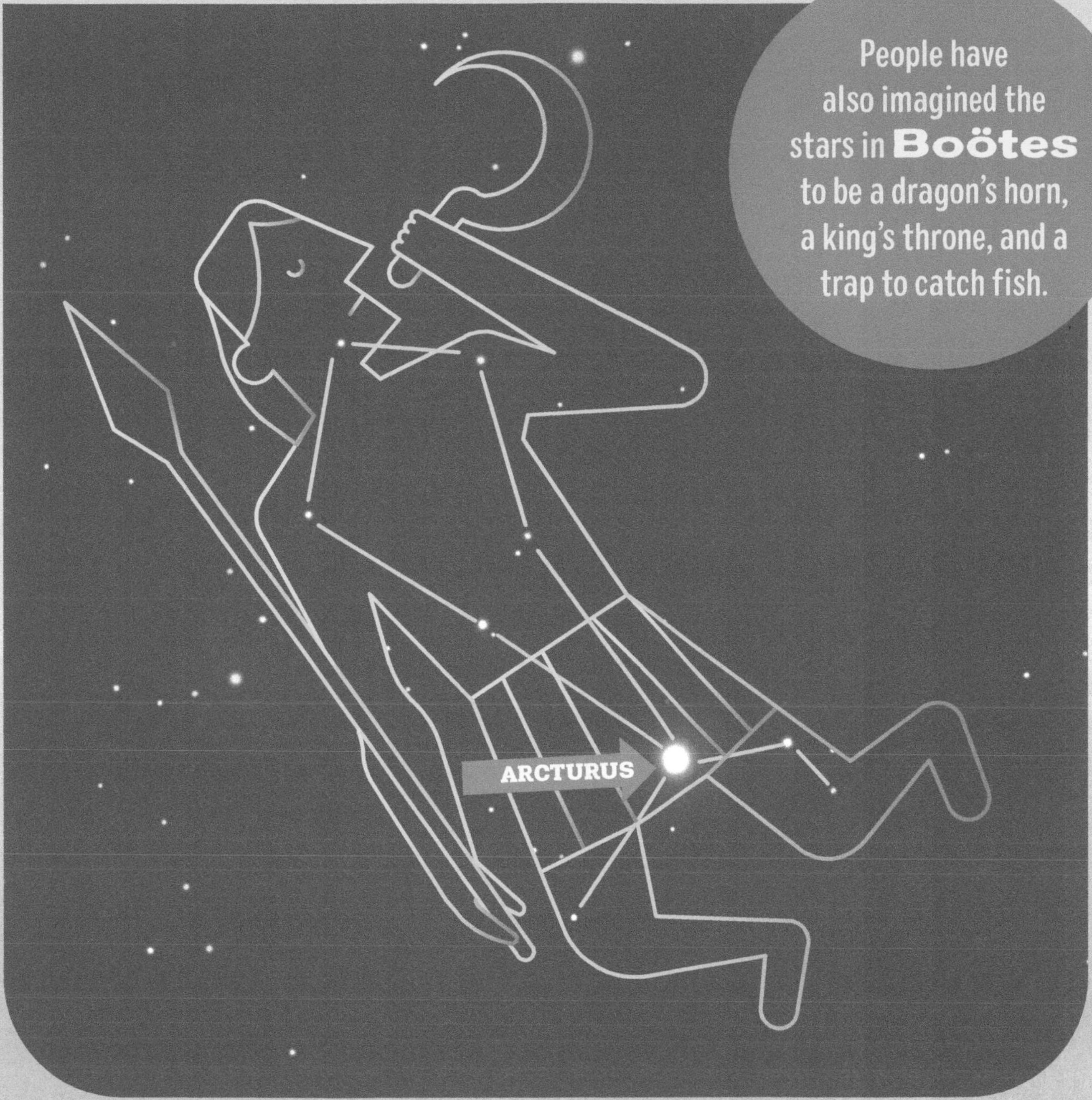

People have also imagined the stars in **Boötes** to be a dragon's horn, a king's throne, and a trap to catch fish.

ARCTURUS

THE MAIDEN OF MANY NAMES

There are many stories about Virgo. In some stories, Virgo is the goddess of grain. In other stories, Virgo is the goddess of justice. Virgo is sometimes also said to be **Persephone**, the goddess of spring.

Virgo

INTERMEDIATE ★★☆

Virgo, the maiden. Virgo is one of the largest constellations in the sky.

BEST MONTHS TO VIEW AT 9:00 P.M.: APRIL TO JUNE			
When to look:	APRIL	MAY	JUNE
Where to look:	SE	SE	S
How high to look:	2 fists (20°)	3 fists (30°)	4 fists (40°)

STAR BY STAR

1. Find the Big Dipper (see page 14 to find out how).
2. Identify the handle of the Big Dipper.
3. Imagine a line following the curve of the handle.
4. Extend that imaginary line out the end of the handle.
5. Follow that curve for three fists (30°).
6. Your line will land on a bright star called *Arcturus*.
7. Continue that same curve past Arcturus for another three fists (30°).
8. Your line should land on another bright star named *Spica*. Spica is the brightest star in the constellation Virgo.

Easy Trick: You can remember how to find Spica by the saying, "Arc to **Arcturus** and speed on to Spica!"

SPICA

Cancer

ADVANCED ★★★

Cancer, the crab. Cancer is one of the faintest constellations in the sky.

BEST MONTHS TO VIEW AT 9:00 P.M.: FEBRUARY TO MAY				
When to look:	FEBRUARY	MARCH	APRIL	MAY
Where to look:	**SE**	**SE**	**S**	**SW**
How high to look:	5 fists (50°)	6 fists (60°)	6 fists (60°)	5 fists (50°)

STAR BY STAR

1. Find Leo (see page 21 to find out how).
2. Identify *Zosma*, the star at the top of the triangle of Leo's tail.
3. Identify *Algieba*, the star located where Leo's back connects with his mane.
4. Imagine a line from Zosma to Algieba.
5. Extend the imaginary line for two fists (20°).
6. Your imaginary line should land in the center of Cancer.

WATCH YOUR TOES!

The Greek hero Hercules was trying to defeat the monster **Hydra**. While Hercules was fighting Hydra, a crab bit his foot. Hercules killed the crab. The goddess **Hera** did not like Hercules, so Hera put the crab in the stars to honor it.

SPRING

In ancient Egypt, the stars of Cancer were seen as a kind of beetle called a **scarab**.

CONSTELLATIONS BEST TO SEE IN SUMMER

Lyra
Cygnus
Sagittarius
Corona Borealis
Ursa Minor
Hercules
Aquila
Scorpius
Libra

Vega was the first star to have its picture taken! The photo was taken in the year **1850**.

Lyra

BEGINNER ★☆☆

Lyra, a harp. The stars in Lyra include one of the brightest stars in the summer sky, named Vega, which makes it relatively easy to find. Lyra looks like a perfect little triangle connected to a **parallelogram** (a slanted rectangle).

BEST MONTHS TO VIEW AT 10:00 P.M.: JUNE TO OCTOBER					
When to look:	JUNE	JULY	AUGUST	SEPTEMBER	OCTOBER
Where to look:	E	E	Straight up!	W	W
How high to look:	4 fists (40°)	6 fists (60°)	9 fists (90°)	7 fists (70°)	4 fists (40°)

STAR BY STAR

1. Face the north-northeast horizon.
2. Imagine a line from the horizon up through the highest point in the sky (known as the **zenith**).
3. The star *Vega* will fall almost exactly on this line for most of the summer months.

MUSIC IN THE STARS

Lyra is the only Greek constellation that is a musical instrument. According to the myth, Lyra was the first harp ever made. The legendary musician **Orpheus** played Lyra with such beauty that he could charm anyone. Zeus, the king of the Greek gods, placed Lyra in the sky to honor Orpheus and his beautiful music.

A SWAN OR A GOD?

The king of the Greek gods, Zeus, pretended to be a swan to meet a woman named Leda. He wanted Leda to love him. Leda had two sets of twin children. Her children included the boys **Castor** and **Pollux**. Castor and Pollux are the twins who make up the constellation Gemini.

Cygnus

BEGINNER ★☆☆

Cygnus, the swan. Cygnus is one of the easiest constellations to find in the summer sky. The main stars in Cygnus form an asterism known as the Northern Cross. If you are at a very dark location, you will see that Cygnus appears to be flying along a faint band of clouds. These clouds are actually the galaxy we live in, which is called the Milky Way.

BEST MONTHS TO VIEW AT 10:00 P.M.: JULY TO OCTOBER				
When to look:	JULY	AUGUST	SEPTEMBER	OCTOBER
Where to look:	E	E	Straight up!	W
How high to look:	4 fists (40°)	7 fists (70°)	9 fists (90°)	6 fists (60°)

STAR BY STAR

1. Find the bright star *Vega* (see page 31 to find out how).
2. Imagine a line connecting Vega to the northeast horizon.
3. Follow this line two and a half fists (25°) northeast from Vega.
4. Your line should land close to the brightest star in Cygnus, named *Deneb*.

DENEB

ALBIREO

SUMMER

The "**star**"

Albireo is at the beak of the swan. Albireo is actually two stars. You can see the two stars that make up Albireo with binoculars.

SCORPION DEFENSE

Sagittarius is a **centaur** in Greek mythology. A centaur is part human and part horse. The constellation Sagittarius is aiming an arrow at the constellation Scorpio, the scorpion. Scorpio was sent to kill the hunter Orion. Sagittarius is protecting Orion from Scorpio.

Sagittarius

BEGINNER ★☆☆

Sagittarius, the archer. Sagittarius is one of the easiest constellations to recognize in the southern sky. The main stars in Sagittarius form the asterism called the **Teapot**. Sagittarius never gets very far above the horizon. You will need a clear horizon to the south to see Sagittarius.

BEST MONTHS TO VIEW AT 10:00 P.M.: JULY TO SEPTEMBER			
When to look:	JULY	AUGUST	SEPTEMBER
Where to look:	SE	S	S
How high to look:	1 fist (10°)	2 fists (20°)	1 fist (10°)

STAR BY STAR

1. Face south.
2. Find the bright star *Altair* (see page 42 to find out how).
3. Measure four fists (40°) to the southwest of Altair.
4. You should land near the asterism of the Teapot.

THE TEAPOT

When Babylonian people saw the stars in **Sagittarius**, they imagined a god with two heads. One of the heads was human, and the other head was a panther!

THE PRINCESS AND THE CROWN

The princess **Ariadne** married the Greek god **Dionysus.** Ariadne wore a crown at her wedding. Dionysus placed the crown among the stars to celebrate their wedding. This crown is now the constellation Corona Borealis.

Corona Borealis

INTERMEDIATE ★★☆

Corona Borealis, the Northern Crown (*corona* is the Latin word for "crown"). This small constellation makes an almost perfect half circle of stars.

BEST MONTHS TO VIEW AT 10:00 P.M.: MAY TO SEPTEMBER					
When to look:	MAY	JUNE	JULY	AUGUST	SEPTEMBER
Where to look:	E	SE	SW	W	W
How high to look:	5 fists (50°)	7 fists (70°)	7 fists (70°)	5 fists (50°)	3 fists (30°)

STAR BY STAR

1. Find the bright star *Arcturus* (see page 24 to find out how).
2. Find the bright star *Vega* (see page 31 to find out how).
3. Imagine a line connecting Arcturus and Vega.
4. Measure two fists (20°) from Arcturus along this imaginary line.
5. You should land on the brightest star in Corona Borealis, *Alphecca*.

ALPHECCA

People have also imagined the stars in **Corona Borealis** were an eagle's nest, a flower garden, a bear den, and a circle of elders.

Ursa Minor

INTERMEDIATE ★★☆

Ursa Minor, the little bear. Because the stars in Ursa Minor form the shape of a serving spoon, people also call it the Little Dipper. Ursa Minor is important because it shows you where the North Star is.

BEST MONTHS TO VIEW AT 10:00 P.M.: MARCH TO OCTOBER								
When to look:	MARCH	APRIL	MAY	JUNE	JULY	AUGUST	SEPTEMBER	OCTOBER
Where to look:	N	N	N	N	N	N	N	N
How high to look:	3 fists (30°)	4 fists (40°)	5 fists (50°)	5 fists (50°)	5 fists (50°)	4 fists (40°)	4 fists (40°)	3 fists (30°)

STAR BY STAR

1. Face the northern horizon.
2. Find Ursa Major, also called the Big Dipper (see page 14 to find out how).
3. Identify the Pointer Stars in the Big Dipper, *Merak* and *Dubhe*.
4. Imagine a line connecting Merak to Dubhe.
5. Extend that imaginary line by three fists (30°).
6. Your imaginary line should land near the North Star.

THE BEAR AND HER SON

The stories of Ursa Minor and Ursa Major go together. The **nymph Callisto** had been turned into a bear. Her son, **Arcus**, did not know that the bear was really his mother, and he tried to kill her. The king of the Greek gods, Zeus, saved Callisto. Zeus had Callisto and Arcus delivered to the sky to protect them. Callisto is Ursa Major. Arcus is Ursa Minor.

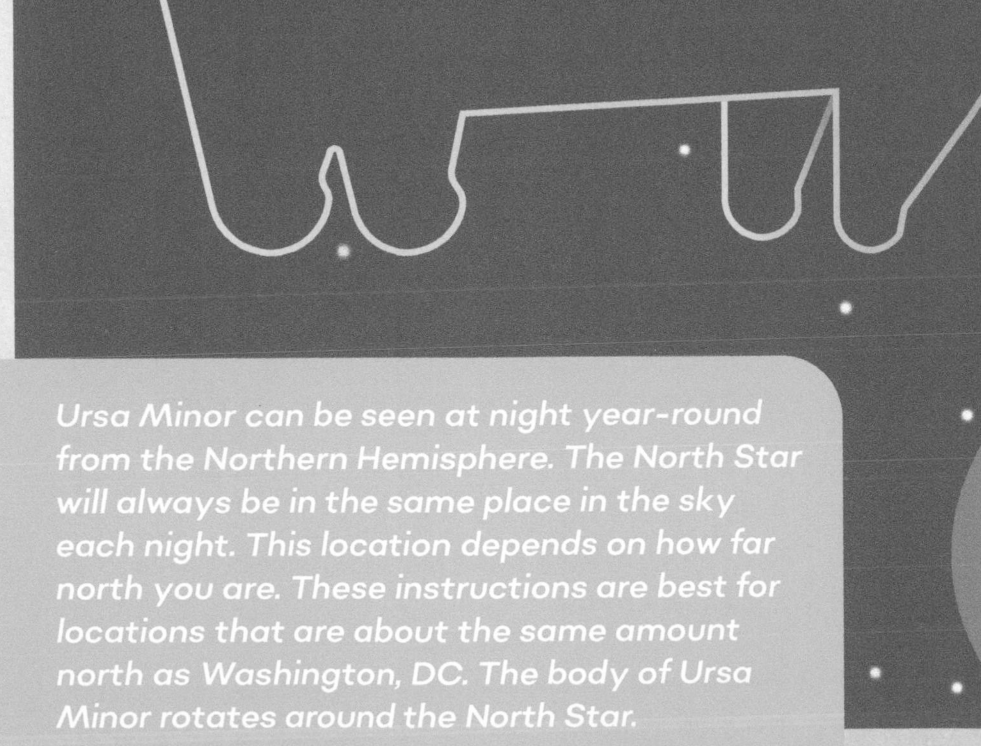

Ursa Minor can be seen at night year-round from the Northern Hemisphere. The North Star will always be in the same place in the sky each night. This location depends on how far north you are. These instructions are best for locations that are about the same amount north as Washington, DC. The body of Ursa Minor rotates around the North Star.

The word "**arctic**" comes from the Greek word *arktos*, which means "**bear**."

THE GREAT HERO

According to Greek mythology, the goddess Hera did not like Hercules. When Hercules was born, she sent snakes to kill him. However, even as a baby Hercules was so strong that he was able to kill the snakes.

Hercules

INTERMEDIATE ★★☆

Hercules, the hero. The main stars in Hercules form a **trapezoid** (a square with only two sides parallel to each other). This trapezoid is an asterism called the **Keystone**.

BEST MONTHS TO VIEW AT 10:00 P.M.: MAY TO SEPTEMBER					
When to look:	MAY	JUNE	JULY	AUGUST	SEPTEMBER
Where to look:	E	E	Straight up!	W	W
How high to look:	3 fists (30°)	5 fists (50°)	9 fists (90°)	6 fists (60°)	4 fists (40°)

STAR BY STAR

1. Find the bright star *Vega* (see page 31 to find out how).
2. Find the bright star *Arcturus* (see page 24 to find out how).
3. Imagine a line connecting Vega and Arcturus.
4. The Keystone asterism will fall on this line between Vega and Arcturus.

Hercules once killed a dragon named **Ladon**. This dragon is a constellation named **Draco**. Draco is near Hercules in the sky.

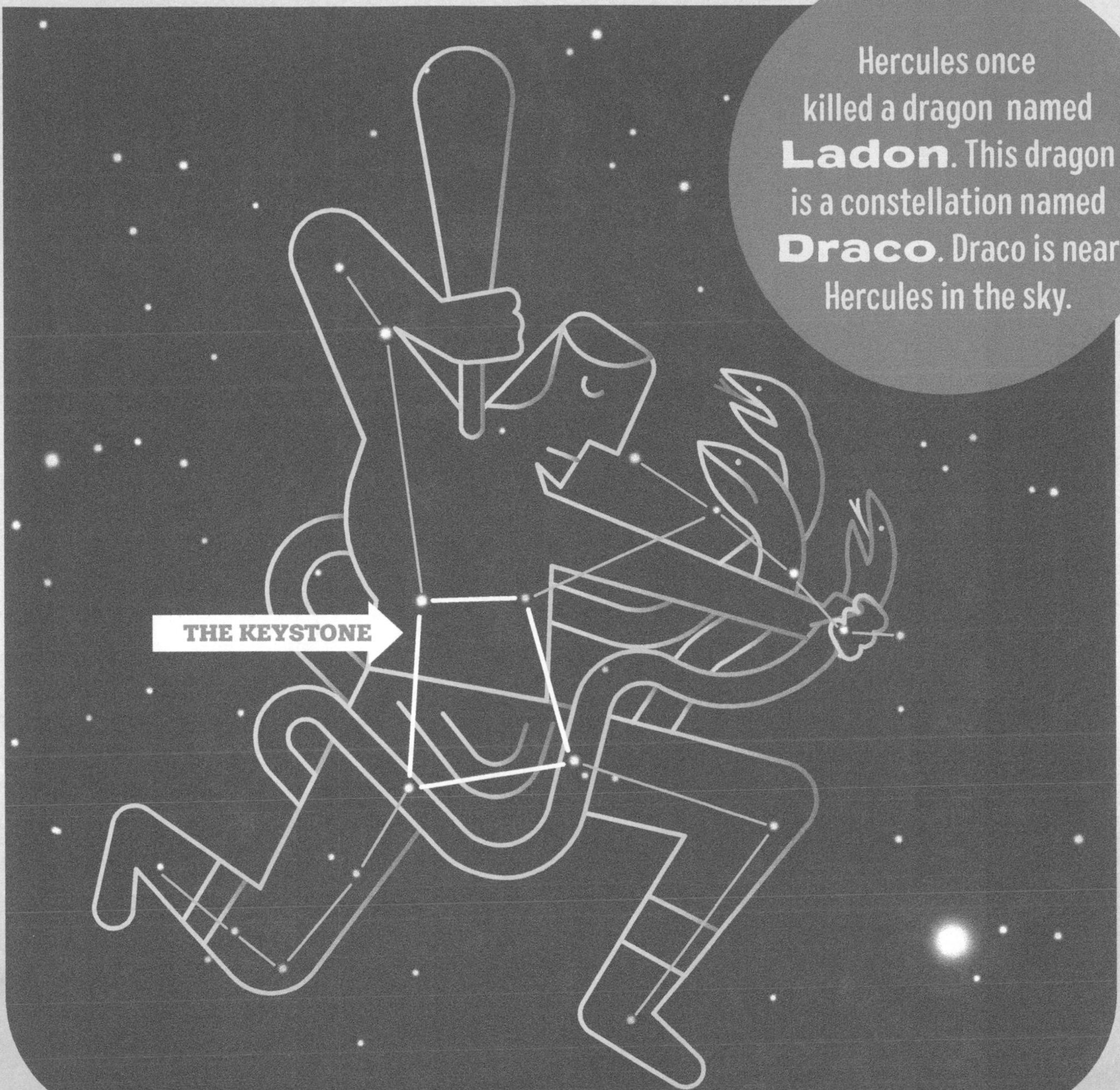

THE KEYSTONE

THUNDERBOLTS FROM THE GODS

Aquila was a bird belonging to the king of the Greek gods, Zeus. Zeus used Aquila to carry thunderbolts to Earth. Aquila also carried a boy named **Ganymede** from Earth to the heavens. Ganymede served water to Zeus. Aquila and the constellation Ganymede (called Aquarius) are near each other in the sky.

Aquila

INTERMEDIATE ★★☆

Aquila, the eagle. Aquila is an important summer constellation. The brightest star in Aquila is named Altair. The bright stars Altair, Vega (in Lyra), and Deneb (in Cygnus) form an asterism known as the **Summer Triangle**.

BEST MONTHS TO VIEW AT 10:00 P.M.: JULY TO OCTOBER				
When to look:	JULY	AUGUST	SEPTEMBER	OCTOBER
Where to look:	E	SE	SW	SW
How high to look:	3 fists (30°)	5 fists (50°)	5 fists (50°)	4 fists (40°)

STAR BY STAR

1. Find the bright star *Vega* (see page 32 to find out how).
2. Connect a line from Vega through the bottom corner of the parallelogram in Lyra.
3. Follow this line for three fists (30°).
4. Your line should land close to the brightest star in Aquila, named *Altair*.

ALTAIR

People have also imagined that the bright stars in **Aquila** were a drum, three soldiers, or a campfire.

PROTECTOR OF ANIMALS

One day, the great hunter Orion said he would kill all of the wild animals on Earth. This made the Greek goddess **Gaia** very angry. Gaia sent a giant scorpion to attack Orion. The battle between Orion and the scorpion impressed the king of the gods, Zeus. Zeus gave the scorpion a place of honor in the sky. Orion and Scorpius cannot be seen in the sky at the same time.

Scorpius

INTERMEDIATE ★★☆

Scorpius, the scorpion. The main stars in Scorpius actually look like a scorpion! The bright star **Antares** will stand out low on the southern horizon.

BEST MONTHS TO VIEW AT 10:00 P.M.: JUNE TO AUGUST			
When to look:	JUNE	JULY	AUGUST
Where to look:	SE	S	SW
How high to look:	2 fists (20°)	3 fists (30°)	2 fists (20°)

STAR BY STAR

1. Face south.
2. Look for the brightest star low on the Southern Horizon. This star might even look a little red to you! This star is called *Antares*.
3. Antares will be about 2 fists above the horizon if you are as far north as Maine, and about 4 fists above the horizon if you are as far south as Florida.
4. The main body of Scorpius looks like a fishhook extending two fists (20°) to the left of Antares.

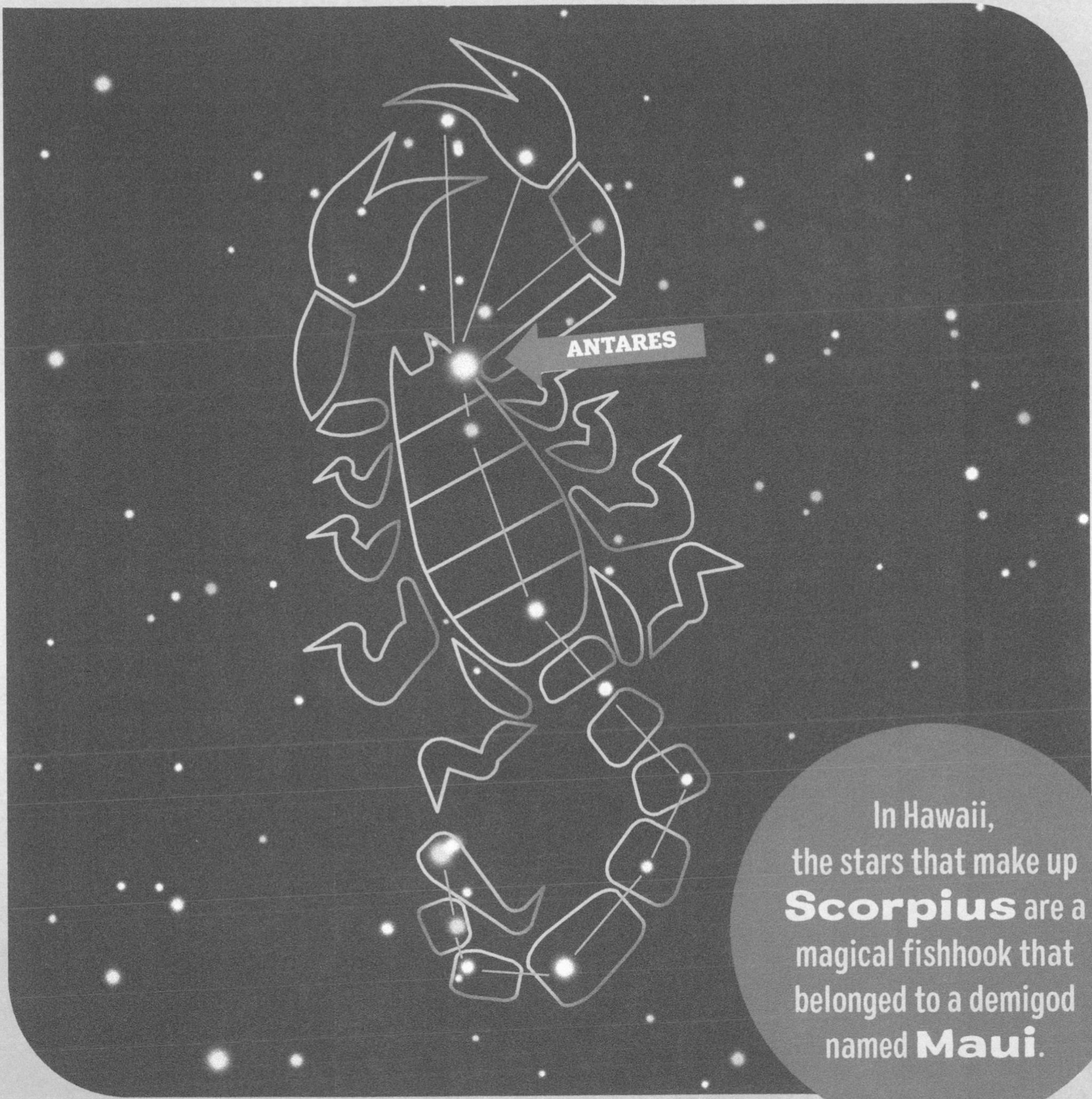

ANTARES

In Hawaii, the stars that make up **Scorpius** are a magical fishhook that belonged to a demigod named **Maui**.

BALANCE AND ORDER

In Greek mythology there was a goddess of law, order, and justice named **Dike**. Dike held balancing scales to determine what was fair. Her balancing scales are represented in the constellation Libra.

Libra

ADVANCED ★★★

Libra, the weighing scales. Libra is a small constellation. The main stars in Libra look like a squished diamond.

BEST MONTHS TO VIEW AT 10:00 P.M.: MAY TO JULY			
When to look:	MAY	JUNE	JULY
Where to look:	SE	S	SW
How high to look:	2 fists (20°)	3 fists (30°)	3 fists (30°)

STAR BY STAR

1. Face south.
2. Find the bright star *Spica* (see page 24 to find out how).
3. Find the bright star *Antares* (see page 44 to find out how).
4. Imagine a line connecting Spica and Antares.
5. Libra will fall in the middle of this imaginary line.

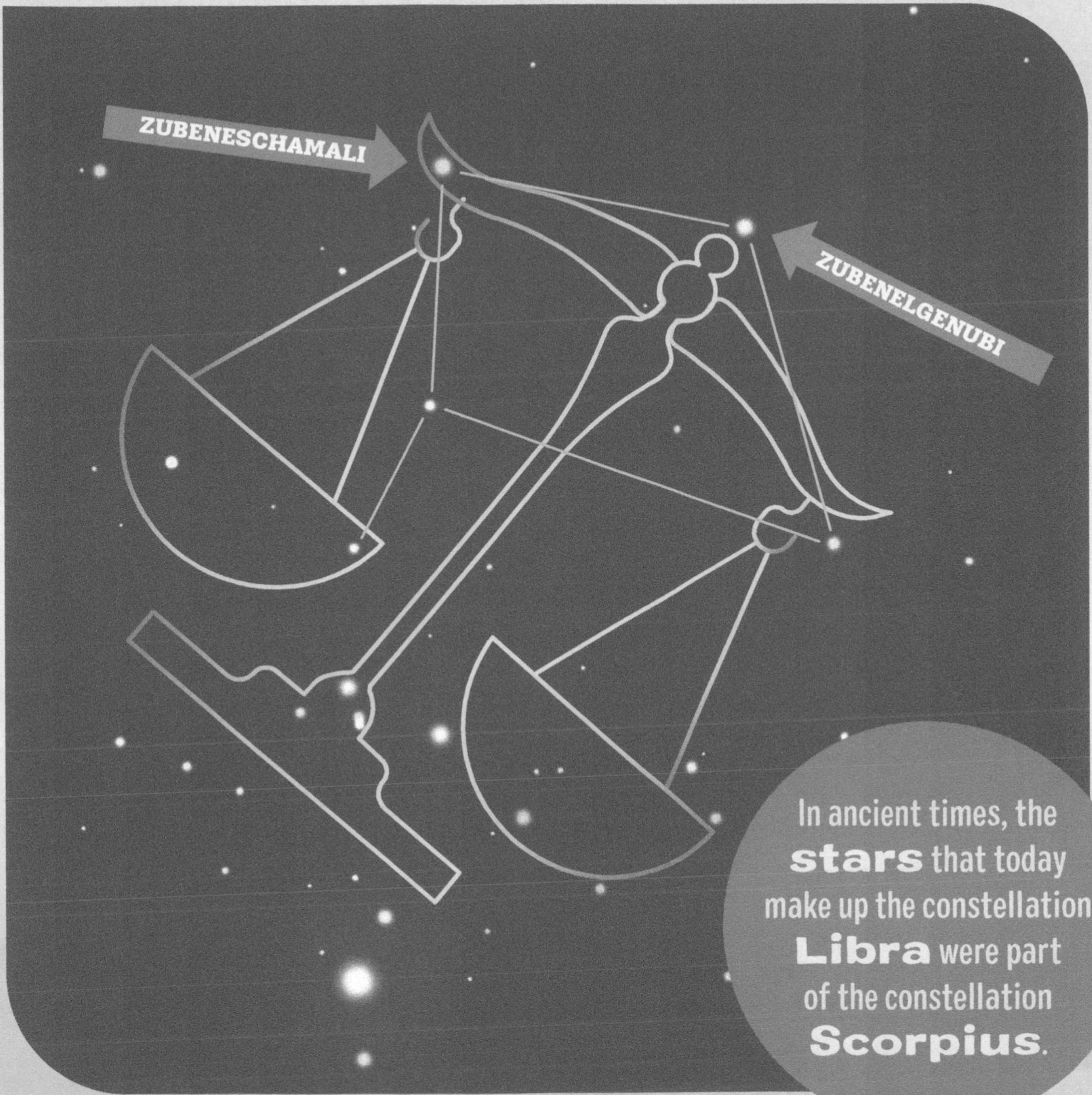

ZUBENESCHAMALI

ZUBENELGENUBI

In ancient times, the **stars** that today make up the constellation **Libra** were part of the constellation **Scorpius**.

CONSTELLATIONS BEST TO SEE IN FALL

Cassiopeia
Pegasus
Aries
Capricorn
Aquarius
Pisces

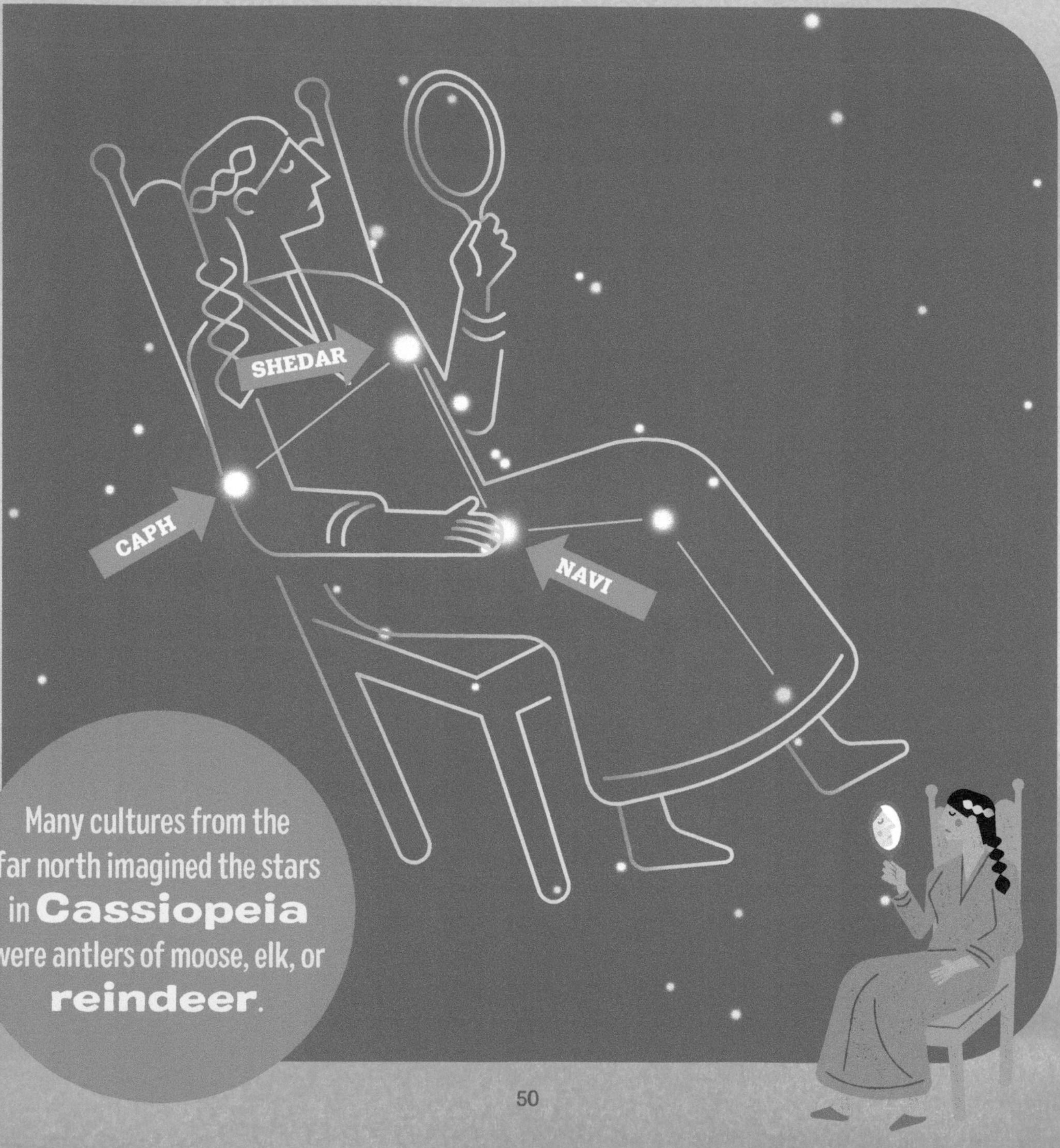

Many cultures from the far north imagined the stars in **Cassiopeia** were antlers of moose, elk, or **reindeer**.

Cassiopeia BEGINNER ★☆☆

Cassiopeia, the proud queen. The main stars in Cassiopeia form a giant W shape that stands out in the northern sky. Cassiopeia can be seen at night year-round from the Northern Hemisphere.

BEST MONTHS TO VIEW AT 9:00 P.M.: SEPTEMBER TO FEBRUARY						
When to look:	SEPTEMBER	OCTOBER	NOVEMBER	DECEMBER	JANUARY	FEBRUARY
Where to look:	NE	NE	N	N	NW	NW
How high to look:	3 fists (30°)	5 fists (50°)	7 fists (70°)	6 fists (60°)	5 fists (50°)	4 fists (40°)

STAR BY STAR

1. Face the northern horizon.
2. Find Ursa Minor (see page 38 to find out how).
3. Identify the North Star.
4. Imagine a circle around the North Star that extends the length of three fists (30°).
5. Pretend that this giant circle is a clock face.
6. Look for Cassiopeia on the imaginary clock!

Cassiopeia was a queen who thought that she was very pretty. One day, she bragged that she was prettier than the daughters of the god of the ocean, **Poseidon**. This made Poseidon angry. He sent a sea monster to attack Cassiopeia's kingdom. As punishment, Cassiopeia was placed in the stars, where she is upside down during half of the year.

THE GREAT HERO

Pegasus was born from the blood of the **gorgon Medusa**. In Greek stories, water would come from the ground wherever Pegasus struck his hoof. After helping a hero defeat a monster, Pegasus lived with the god Zeus. Zeus rewarded Pegasus by giving him a spot in the stars.

Pegasus

BEGINNER ★☆☆

Pegasus, the winged horse. Four of the bright stars in Pegasus form the asterism the **Great Square**.

BEST MONTHS TO VIEW AT 9:00 P.M.: SEPTEMBER TO DECEMBER				
When to look:	SEPTEMBER	OCTOBER	NOVEMBER	DECEMBER
Where to look:	E	E	S	W
How high to look:	3 fists (30°)	6 fists (60°)	7 fists (70°)	5 fists (50°)

STAR BY STAR

1. Find the constellation Cassiopeia (see page 51 to find out how).
2. Identify the stars *Navi* and *Shedar* in Cassiopeia.
3. Imagine a line going from Navi through Shedar.
4. Follow that line for about five fists (50°).
5. Your line should land right on Pegasus!

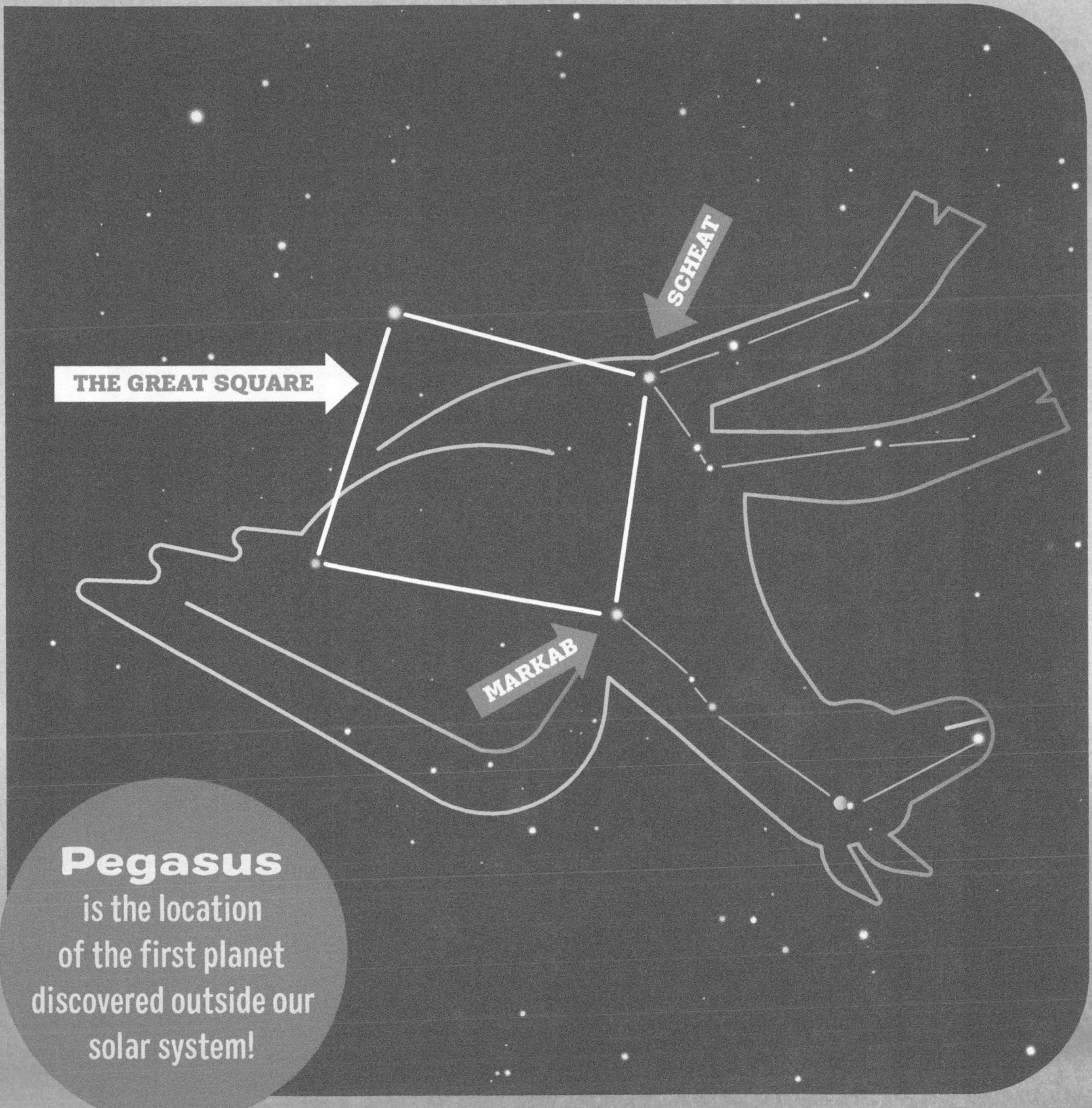

THE GREAT SQUARE

SCHEAT

MARKAB

Pegasus
is the location
of the first planet
discovered outside our
solar system!

THE GOLDEN FLEECE

Aries was a beautiful ram with golden fleece. After the ram was sacrificed to the Greek god Zeus, the golden fleece was protected by a dragon. Zeus gave Aries a place of honor in the night sky.

Aries

INTERMEDIATE ★★☆

Aries, the ram. Aries only has a few stars bright enough to see from most locations. This constellation will look like a crooked line with three main stars.

BEST MONTHS TO VIEW AT 9:00 P.M.: OCTOBER TO JANUARY				
When to look:	OCTOBER	NOVEMBER	DECEMBER	JANUARY
Where to look:	E	SE	S	W
How high to look:	3 fists (30°)	7 fists (70°)	7 fists (70°)	5 fists (50°)

STAR BY STAR

1. Find the constellation Cassiopeia (see page 51 to find out how).
2. Find the stars *Caph* and *Shedar* on the right side of the "W" shape.
3. Imagine a line connecting Caph and Shedar.
4. Follow that imaginary line for four fists (40°).
5. Your imaginary line will land on the constellation Aries!

HAMAL

The stars in Aries have also been seen as a **porpoise**.

ESCAPE FROM A MONSTER

In some stories Capricorn was a god named **Pan**. Pan had the horns and legs of a goat. One day, Pan escaped a monster by jumping into a river. When he jumped into the river, he got the tail of a fish.

Capricorn

ADVANCED ★★★

Capricorn, the sea goat. Capricorn is one of the faintest constellations. With patience and practice you can find it! The main stars in Capricorn are shaped like a squished triangle.

BEST MONTHS TO VIEW AT 9:00 P.M.: SEPTEMBER TO NOVEMBER			
When to look:	SEPTEMBER	OCTOBER	NOVEMBER
Where to look:	SE	S	SW
How high to look:	3 fists (30°)	3 fists (30°)	3 fists (30°)

STAR BY STAR

1. Face south.
2. Find the bright star *Altair* (see page 42 to find out how).
3. Find the bright star *Vega* (see page 31 to find out how).
4. Imagine a line from Altair through Vega.
5. Continue the imaginary line for three fists (30°).
6. Your line will land on Capricorn.

The brightest star in Capricorn is named ***Algiedi***, which means "**goat**" in Arabic.
ALGIEDI

SNATCHED BY AN EAGLE

There was a boy in Greek stories named Ganymede. Ganymede was taken by a great eagle to the god Zeus. Ganymede brought cups of water to the gods who lived on Mount Olympus. The eagle who took Ganymede to Zeus is the constellation Aquila. Aquarius means "water carrier."

Aquarius

ADVANCED ★★★

Aquarius, the water carrier. Aquarius does not have any bright stars. It will take practice and patience to find this constellation!

BEST MONTHS TO VIEW AT 9:00 P.M.: SEPTEMBER TO NOVEMBER			
When to look:	SEPTEMBER	OCTOBER	NOVEMBER
Where to look:	SE	S	SW
How high to look:	3 fists (30°)	4 fists (40°)	3 fists (30°)

STAR BY STAR

1. Find the Great Square of Pegasus (see page 53 to find out how).
2. Identify the stars *Scheat* and *Markab*.
3. Imagine a line connected from Scheat through Markab.
4. Follow that imaginary line for two fists (20°).
5. Your line should land in Aquarius.

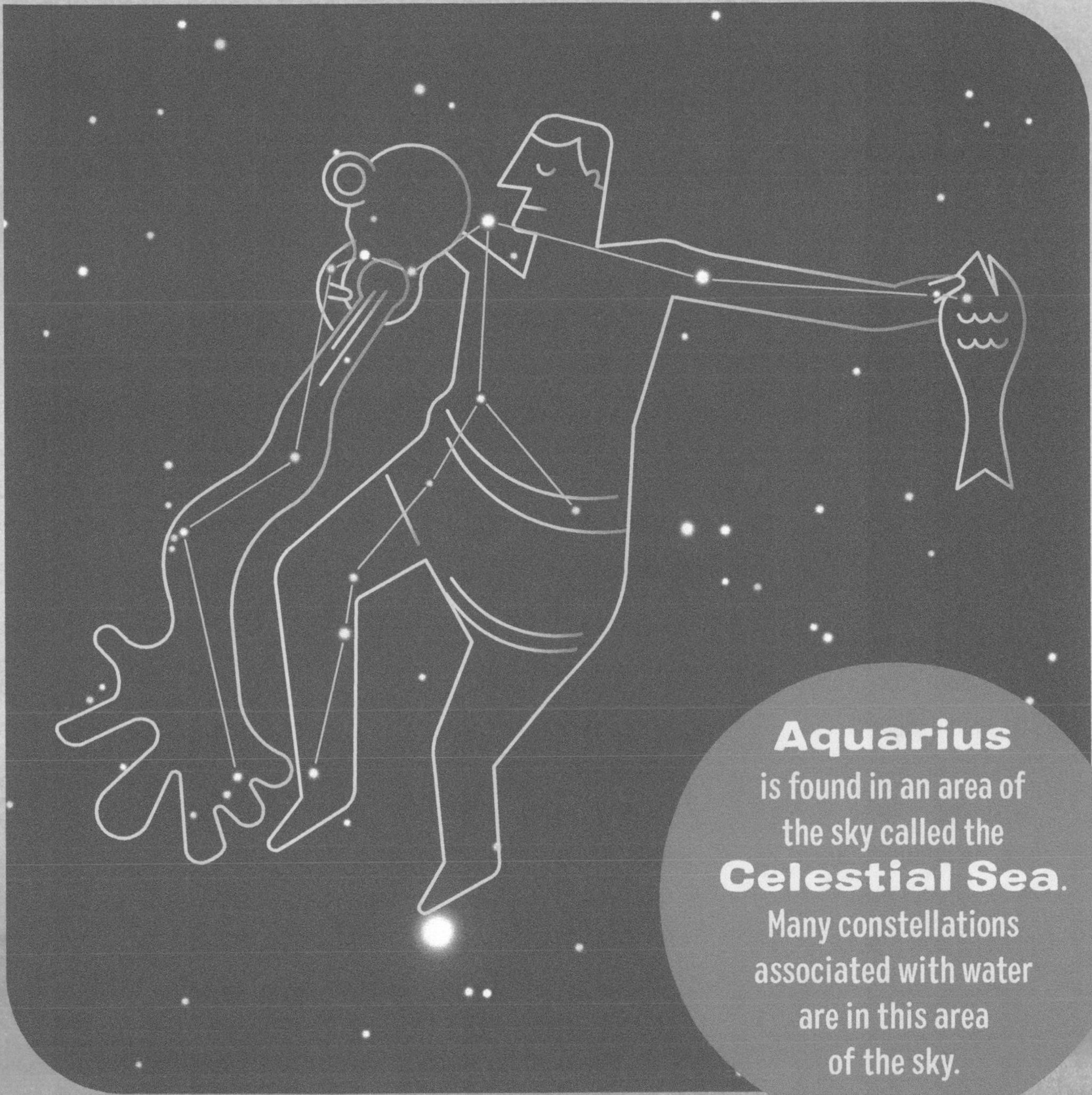

Aquarius is found in an area of the sky called the **Celestial Sea**. Many constellations associated with water are in this area of the sky.

THE MOTHER AND SON

When a scary monster was attacking the Greek gods, the goddess **Aphrodite** and her son jumped into a river to escape. They turned into fish! The goddess and her son tied a ribbon to each other so they would not be separated. These fish are connected by a ribbon in the constellation Pisces.

Pisces

ADVANCED ★★★

Pisces, the two fish. Pisces is a large constellation, but it does not have any bright stars. It will take practice and patience to find this constellation!

BEST MONTHS TO VIEW AT 9:00 P.M.: OCTOBER TO JANUARY				
When to look:	OCTOBER	NOVEMBER	DECEMBER	JANUARY
Where to look:	E	S	SW	W
How high to look:	3 fists (30°)	6 fists (60°)	5 fists (50°)	3 fists (30°)

STAR BY STAR

1. Find the Great Square in the constellation Pegasus (see page 53 to find out how).
2. One of the fish in Pisces is roughly one fist (10°) to the east of the Great Square.
3. One fish in Pisces is roughly one fist (10°) to the south of the Great Square.
4. The brightest star in Pisces is roughly two fists (20°) to the southeast of the Great Square.

ALRESCHA

The brightest star in Pisces is named *Alrescha*, which means "**the cord**" and refers to the ribbon that ties the fish together.

CONSTELLATIONS BEST TO SEE IN

WINTER

Orion
Taurus
Canis Major
Gemini

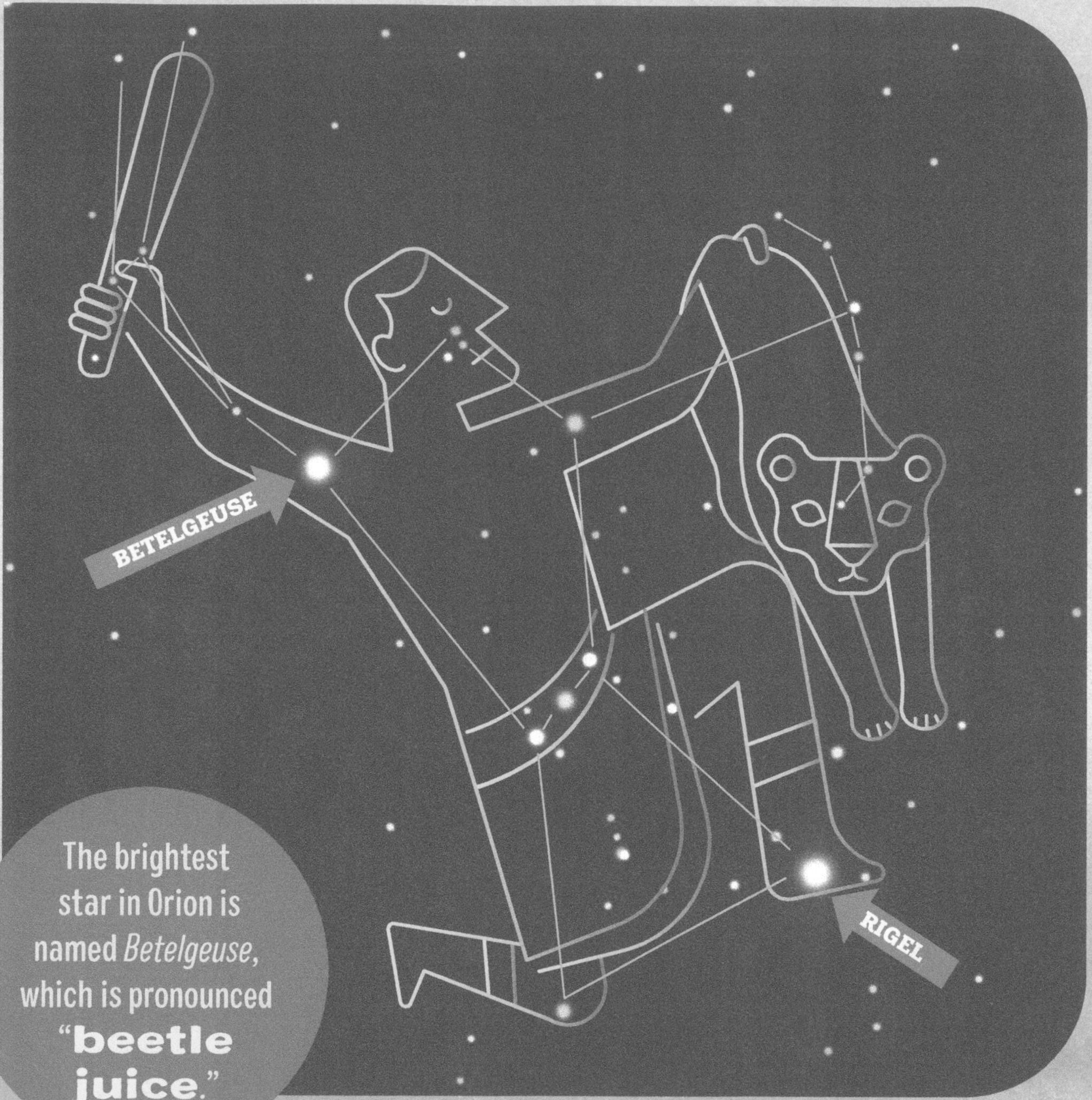

The brightest star in Orion is named *Betelgeuse*, which is pronounced "**beetle juice**."

Orion

BEGINNER ★ ☆ ☆

Orion, the hunter. Orion is one of the best-known constellations. Orion is shaped like a rectangle with the middle squeezed in by his belt.

BEST MONTHS TO VIEW AT 8:00 P.M.: JANUARY TO MARCH			
When to look:	JANUARY	FEBRUARY	MARCH
Where to look:	SE	S	SW
How high to look:	3 fists (30°)	5 fists (50°)	4 fists (40°)

STAR BY STAR

1. Face the southern horizon.
2. Orion will span roughly three to six fists (30° to 60°) above the horizon.
3. Look for the three medium-bright stars of Orion's belt.
4. Orion's shoulders and knees are not quite one fist (10°) above and below his belt.
5. From a dark location you can even see his dagger hanging from his belt!

OH BOY!

Orion was the son of a poor shepherd. One day, the shepherd had visitors. Even though the shepherd was very poor, he gave the visitors food. The shepherd did not know the visitors were actually gods. The grateful gods gave the shepherd a wish. He wished for a son. That son was Orion.

TWO STORIES IN ONE

The Greek god Zeus fell in love with a princess named **Europa**. He changed himself into a beautiful white bull and carried the princess away. The constellation Taurus represents Zeus as this bull. The constellation Taurus also includes the star cluster called the Pleiades. In Greek stories, the Pleiades were seven sisters. The sisters were being courted by the hunter Orion. Zeus put the sisters in the sky to keep them away from Orion.

Taurus

BEGINNER ★☆☆

Taurus, the bull. Taurus has a bright V-shaped head and a famous star cluster named the Pleiades on its shoulder.

BEST MONTHS TO VIEW AT 8:00 P.M.: DECEMBER TO MARCH				
When to look:	DECEMBER	JANUARY	FEBRUARY	MARCH
Where to look:	SE	SE	S	SW
How high to look:	4 fists (40°)	6 fists (60°)	7 fists (70°)	5 fists (50°)

STAR BY STAR

1. Find the constellation Orion (see page 65 to find out how).
2. Imagine a line connecting the stars in Orion's belt.
3. Extend that imaginary line roughly three fists (30°) to the west.
4. Your imaginary line should land near the head of Taurus, which is a V shape.
5. Continue your imaginary line for another fist (10°).
6. This additional line should land near the star cluster called the *Pleiades*.

The Pleiades are also known as **Subaru** in Japanese culture. The next time you see a Subaru car, make sure to look at the logo!

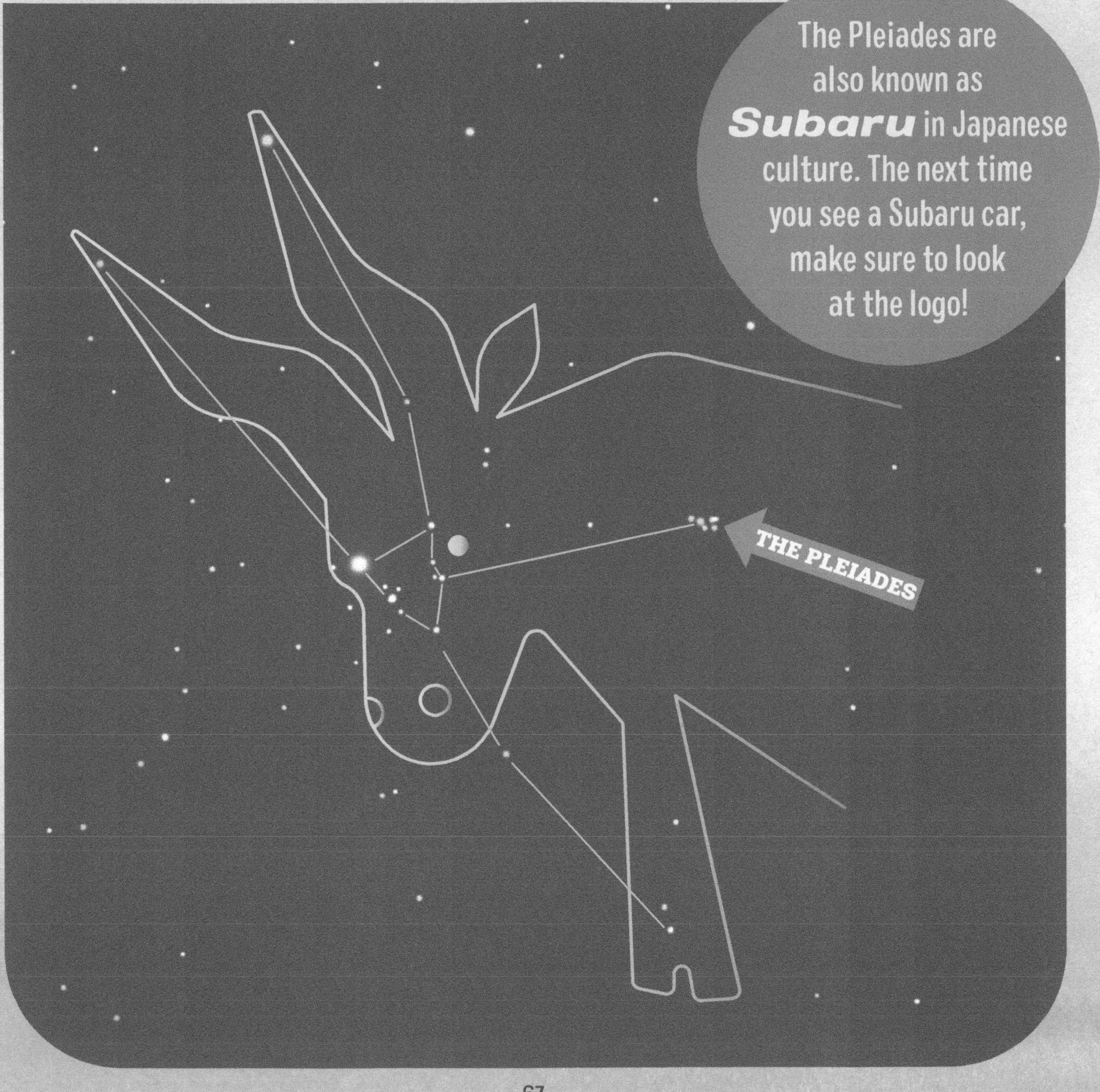

THE PLEIADES

WINTER

ONE FAST DOG

In some stories, Canis Major was a dog that could run really fast. The dog was given as a gift to the Greek god Zeus. Zeus honored the dog by giving him a place in the sky. In another story, Canis Major is the hunting dog of Orion. This is why Canis Major is next to Orion in the sky.

Canis Major

INTERMEDIATE ★★☆

Canis Major, the large dog. Canis Major has the brightest star in the night sky, named Sirius.

BEST MONTHS TO VIEW AT 8:00 P.M.: FEBRUARY TO APRIL			
When to look:	FEBRUARY	MARCH	APRIL
Where to look:	SE	S	SW
How high to look:	3 fists (30°)	3 fists (30°)	3 fists (30°)

STAR BY STAR

1. Face the southern horizon.
2. Find the constellation Orion (see page 65 to find out how).
3. Imagine a line connecting the stars in Orion's belt.
4. Follow that imaginary line about two fists (20°) to the southeast.
5. Your line should land near *Sirius*, the brightest star in Canis Major.

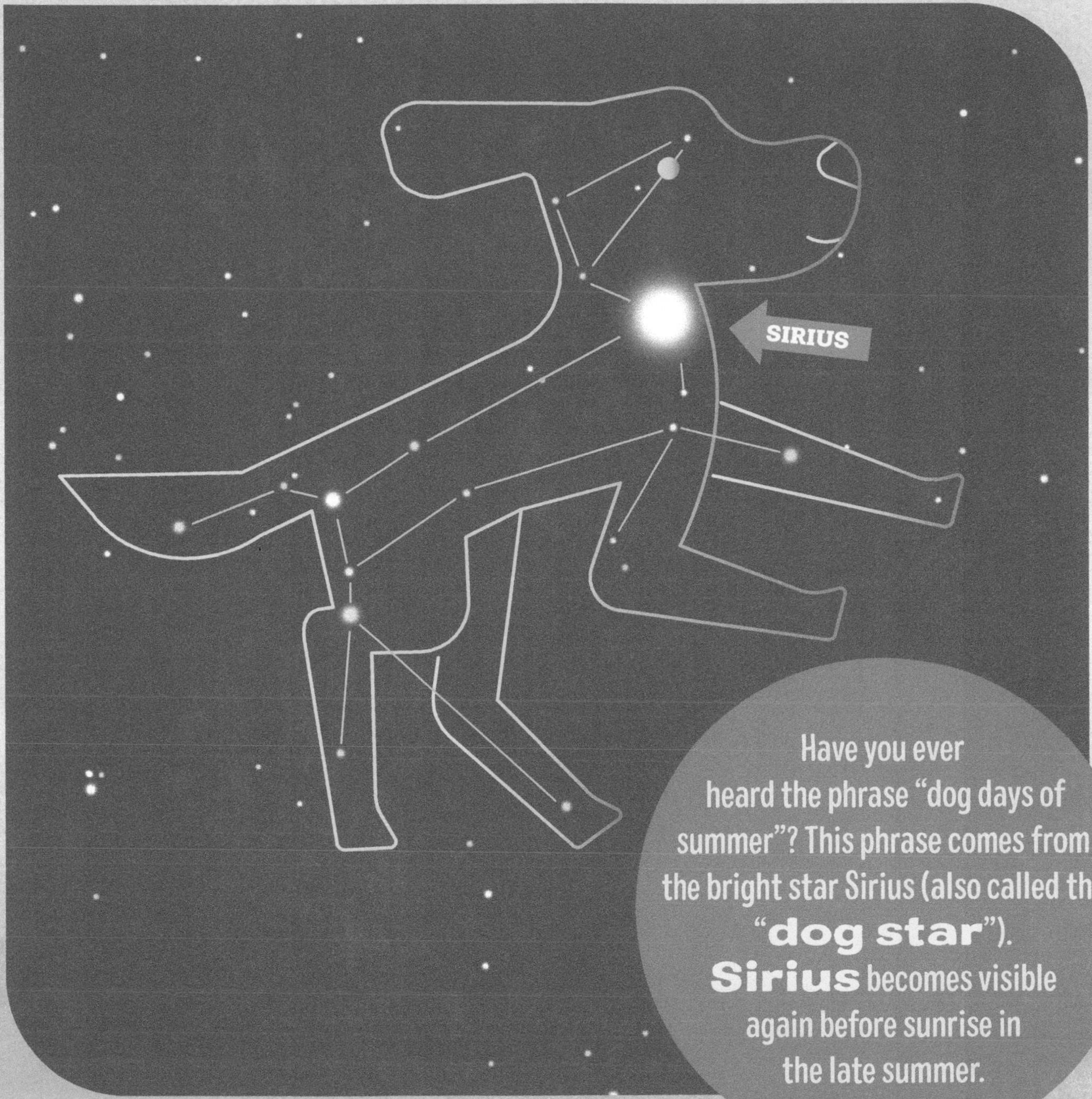

SIRIUS

Have you ever heard the phrase "dog days of summer"? This phrase comes from the bright star Sirius (also called the "**dog star**"). **Sirius** becomes visible again before sunrise in the late summer.

TWINS FOR ETERNITY

In Greek stories, there were twins named Castor and Pollux. Pollux was immortal, but Castor was mortal. When Castor died, Pollux pleaded with the Greek god Zeus. Pollux wanted Zeus to make Castor immortal, too. Zeus united the brothers in the sky.

Gemini

INTERMEDIATE ★★☆

Gemini, the twins. *Gemini* means "twins" in Latin. Two bright stars mark the heads of the twins in this constellation.

BEST MONTHS TO VIEW AT 8:00 P.M.: JANUARY TO APRIL				
When to look:	JANUARY	FEBRUARY	MARCH	APRIL
Where to look:	E	SE	S	SW
How high to look:	4 fists (40°)	7 fists (70°)	7 fists (70°)	6 fists (60°)

STAR BY STAR

1. Find the constellation Orion (see page 65 to find out how).
2. Identify the bright stars *Betelgeuse* and *Rigel.*
3. Imagine a line from Betelgeuse through Rigel.
4. Extend that imaginary line for roughly three fists (30°).
5. Your imaginary line should land near the heads of the twins, the bright stars *Castor* and *Pollux*.

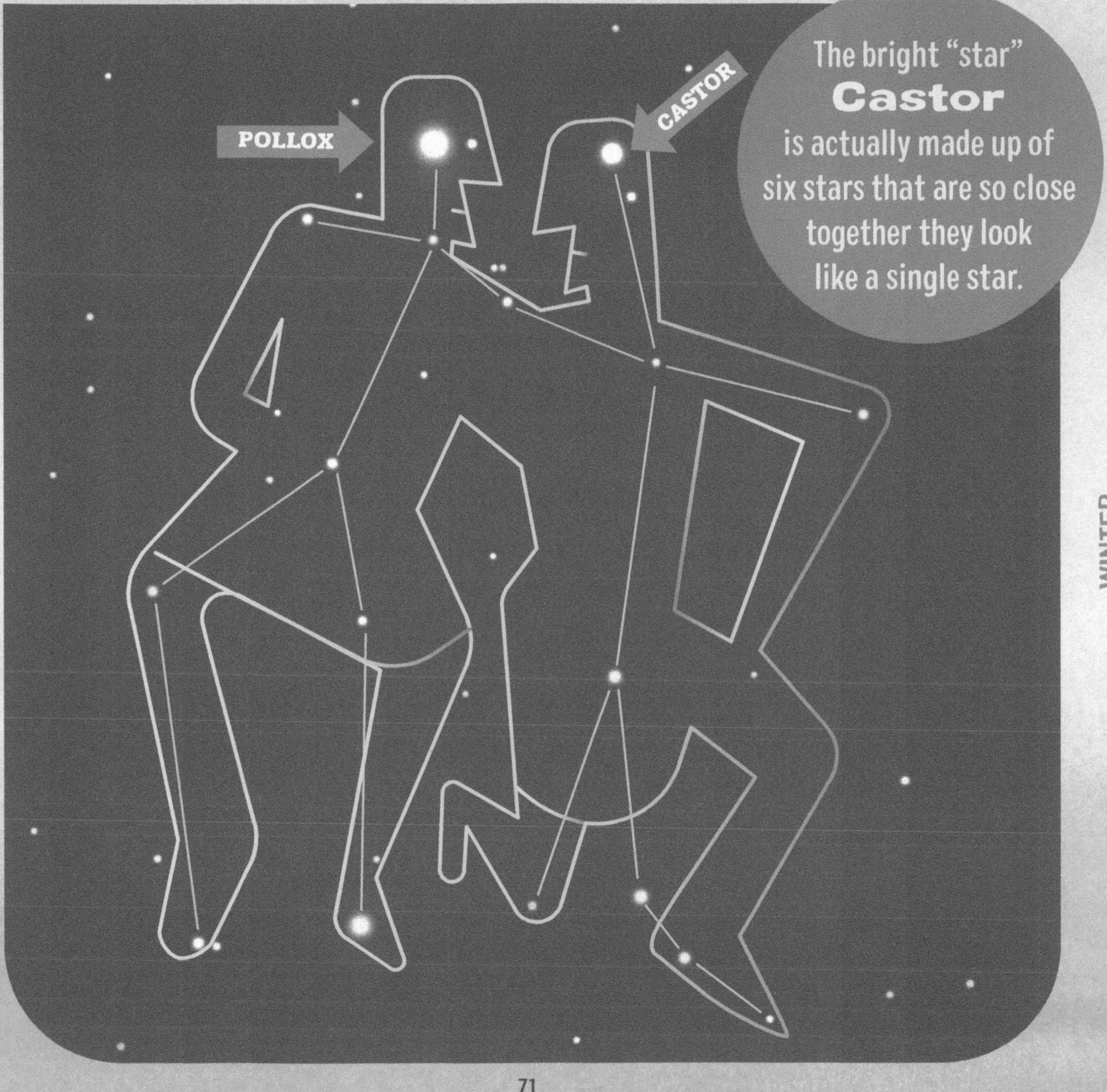

POLLOX

CASTOR

The bright "star"
Castor
is actually made up of
six stars that are so close
together they look
like a single star.

CONGRATULATIONS!

You have now learned the ancient skill of finding your way around the night sky!

This guide to the stars has only scratched the surface. There is so much more to learn! The universe is big (really, *really* big) and full of mysteries. In fact, astronomers have more questions than answers.

Stars, galaxies, black holes, supernova, pulsars, planets, and asteroids are waiting for you to learn about them.

Be curious. Ask questions. Help us all find answers.

Here's to clear skies!

GLOSSARY

Aphrodite: The Greek goddess of love. She and her son turned into fish to escape a monster. The two fish are the constellation Pisces.

Arcus: The son of the nymph Callisto. He is represented by the constellation Ursa Minor.

Ariadne: The princess who married the Greek god Dionysus. The crown she wore at her wedding is represented by the constellation Corona Borealis.

asterism: A pattern of stars that is smaller than a constellation.

Callisto: A nymph who was turned into a bear. Mother of Arcus. She is represented by the constellation Ursa Major.

Castor: One of the twins represented in the constellation Gemini. Castor was mortal. His brother Pollux was immortal.

centaur: A mythological creature that is part human and part horse.

constellation: A pattern of stars in the sky. Today the sky is divided into 88 official constellations.

crescent moon: A phase of the Moon. During a crescent moon only a sliver of the Moon is bright.

degree: A unit of measurement for the size of an angle, often shown as the symbol °.

Dike: The Greek goddess of law, order, and justice. Her balancing scales are the constellation Libra.

Dionysus: The Greek god who married the princess Ariadne.

equator: An imaginary line that goes around the middle of the Earth. This line divides the Earth into two spheres called the Northern and Southern hemispheres.

Europa: A princess whom the Greek god Zeus carried away while in the form of a white bull. This bull is the constellation Taurus.

Gaia: The Greek goddess of the Earth. She sent a giant scorpion to attack the hunter Orion.

Ganymede: A boy who was carried to the Greek god Zeus by the giant bird Aquila. He served water to the gods. Ganymede is represented by the constellation Aquarius.

gorgon: A mythological creature who had snakes for hair. Medusa was a gorgon.

Great Square: An asterism in the constellation Pegasus.

hemisphere: Half of a sphere. The equator divides the Earth into the Northern and Southern hemispheres.

Hera: The queen of the Greek gods and wife of Zeus. She put the constellation Cancer into the sky to honor the crab who fought Hercules.

Hercules: A Greek hero. He defeated the lion Leo as one of his 12 tasks.

Hydra: A water monster who fought Hercules.

Keystone: An asterism in the constellation Hercules.

Medusa: A mythological creature called a gorgon. The winged horse Pegasus was born from her blood.

meteor: A small rock or dust from outer space that flies through Earth's atmosphere, creating streaks of light. They are often called "shooting stars." But they are not stars!

meteor shower: When a large number of meteors fly through Earth's atmosphere during a short time.

new moon: A phase of the Moon when the side of the moon facing the Earth is dark.

nymph: A mythological creature that lives in nature.

orbit: The path that something makes around a star or planet.

Orpheus: A legendary musician in Greek mythology. He played the instrument that is the constellation Lyra.

Pan: The god of the wild in Greek mythology. Pan escaped a monster by jumping into a river and getting a tail. He is shown as the constellation Capricorn.

parallelogram: A type of shape in which opposite sides are the same length and parallel, but the sides can be slanted. It looks like a slanted rectangle.

Persephone: The Greek goddess of spring. In some stories, she is shown by the constellation Virgo.

phase: The shape of the sunlit portion of the Moon that we can see from Earth.

Pleiades: A famous cluster of stars located in the constellation Taurus.

Pollux: One of the twins represented in the constellation Gemini. Pollux was immortal. His brother Castor was mortal.

Poseidon: The Greek god of the ocean. He sent a monster to attack the kingdom ruled by Cassiopeia.

satellite: An object in space that orbits another object that is more massive.

Summer Triangle: A triangle-shaped asterism formed by the stars Altair, Deneb, and Vega.

Teapot: An asterism in the constellation Sagittarius.

trapezoid: A shape that is like a rectangle, but only one pair of opposite sides is parallel.

zenith: The highest point in the sky, directly overhead.

Zeus: The king of the Greek gods. He has a role in many stories related to the constellations.

RESOURCES

Learn to find constellations in the Southern Hemisphere!

You can download your own star map (called a planisphere) of the Southern Hemisphere from this site:

LawrenceHallOfScience.org/sites/default/files/pdfs/starwheels/SouthStarwheel2.pdf

See what the night sky will look like at any time from anywhere on Earth!

You can find planets, constellations, and more. Download a free planetarium program called Stellarium that you can run on a home computer:

Stellarium.org

Learn more about constellations in cultures from around the world!

Aveni, Anthony. *Star Stories: Constellations and People*. New Haven, CT: Yale University Press, 2019.

Ganeri, Anita, and Andy Wilx. *Star Stories: Constellation Tales from around the World*. New York: Running Press Kids, 2019.

Miller, Dorcas S. *Stars of the First People: Native American Star Myths and Constellations*. Boulder, CO: Pruett Publishing Co., 1997.

Want to see a meteor shower?

The American Meteor Society has up-to-date information on upcoming meteor showers.

AMSMeteors.org/meteor-showers/meteor-shower-calendar

Wondering what the next phase of the Moon will be?

You can find Moon phases for any time and date from this site: StarDate.org/nightsky/moon

Want to find out more about light pollution?

Learn how much light pollution from people and cities affects the night sky where you live from this site: LightPollutionMap.info

SKY MAP: SPRING

SKY MAP: SUMMER

SKY MAP: FALL

N
Ursa Major
NE
NW
Ursa Minor
Cassiopeia
Bootes
Aries
Corona Borealis
Hercules
E
Cygnus
Lyra
W
Pisces
Pegasus
Aquila
Aquarius
Capricornus
Sagittarius
SE
SW
S

SKY MAP: WINTER

INDEX

ABOUT THE AUTHOR

Kelsey Johnson teaches students both inside and outside of the classroom, using astronomy as a "gateway" science to nurture curiosity and support science literacy. She is a professor of astronomy at the University of Virginia and founding director of the award-winning Dark Skies, Bright Kids program. She has won numerous awards for her research, teaching, and promotion of science literacy. Her TED talk on the importance of dark skies has over 1.5 million views, and her writing has appeared in *Scientific American*, the *Washington Post*, the *New York Times*, *Ms. Magazine,* and the *Chronicle of Higher Education*. She earned her BA in physics from Carleton College and her MS and PhD in astrophysics from the University of Colorado. She lives in rural Virginia with her family, including two cats and two very large dogs—one of whom is named after the constellation Lyra.

CPSIA information can be obtained
at www.ICGtesting.com
Printed in the USA
JSHW050820030920
7611JS00005B/5

9 781646 119684